Columbia

The Forgotten Supercontinent

JOHN S. PRITCHETT

TABLE OF CONTENTS

1. The First World Assembled ______ 1
2. Fragments of the Ancient Earth _ 13
3. A Hotter, Younger Planet ______ 22
4. The Collisions of Continents ____ 32
5. Mountains That No Longer Exist 44
6. Fire Beneath the Crust _________ 56
7. Diamonds from the Deep ______ 66
8. Metals of Ancient Collisions ____ 77
9. Mapping the Fragments ________ 87
10. The Breakup of Columbia _____ 99
11. Clues Hidden in Today's Continents ________________________________ 110
12. Following the Ancient Scars __ 122

13. Where the Next Discoveries May Lie ______________________________131
14. The Supercontinent Cycle _____141
15. Treasures Written in Stone __ 154

Chapter 1
The First World Assembled

It doesn't look like much at first.

A stretch of exposed rock. Hard, worn, unremarkable. In places like northern Canada, West Africa, or parts of Australia, the land opens into broad, quiet expanses of stone. No sharp peaks. No dramatic cliffs. Just flat, ancient ground, broken by lakes and low ridges.

It feels still. Permanent.

But that surface is older than almost anything we can easily comprehend. Some of these rocks formed more than two billion years ago. They have survived processes that erased entire mountain ranges, opened and closed oceans, and rearranged the continents again and again.

They are not just old.

They are remnants of a world that no longer exists.

The Earth beneath our feet appears stable, but the continents preserve a far older, more restless history. Across billions of years, mountains have risen and vanished, oceans have opened and closed, and landmasses have come together into vast supercontinents before breaking apart again.

That cycle has repeated many times.

One of the earliest of those worlds was Columbia.

Nearly two billion years ago, based on available evidence, the planet looked nothing like it does today. The continents we recognize had not yet formed in their present arrangement. Instead, Earth's surface was scattered with smaller fragments of ancient crust known as cratons, the oldest and most stable pieces of continental material.

These fragments were not large continents. They were the foundations of continents.

Today, they are embedded within the landmasses we know. The Canadian Shield in North America, the ancient rocks of West Africa, Siberia, and Australia all preserve pieces of these early building blocks. Though now separated by vast oceans, geological evidence indicates they were once joined.

The separation we see today is not how things began.

Over immense spans of time, tectonic plates carried these cratons across the surface of the Earth. The motion was slow, about as fast as your fingernails grow. In a single year, the change would be impossible to notice. Over a human lifetime, the movement would barely shift the distance between two points.

But over one hundred million years, that same motion carries continents across oceans.

Slow movement becomes immense change.

Gradually, these fragments began to converge.

When continents meet, the forces involved reshape the planet on a continental scale. The crust thickens.

Layers of rock compress and fold. Mountain ranges rise, not over years or centuries, but over tens of millions of years. What begins as slow motion becomes transformation.

Across hundreds of millions of years, repeated collisions appear to have stitched these cratons together.

Piece by piece, a single landmass emerged.

Evidence suggests this landmass was Columbia.

For a time, much of Earth's continental crust appears to have existed as one connected world. Its exact shape is still debated, but the evidence for its existence is strong. Regions now separated by oceans share matching rock ages, aligned mountain belts, and mineral systems formed during the same events.

These are unlikely coincidences.

They are connections preserved in stone.

The assembly of Columbia likely transformed the planet. Enormous mountain ranges rose where continents collided, rivaling the greatest on Earth today. Their peaks may have stretched for thousands of miles, forming barriers that shaped climate, erosion, and the flow of water across the land.

Those mountains are gone now.

But when they formed, they changed everything around them.

Deep within the crust, the pressure and heat of collision altered rock, creating new structures that would endure long after the surface eroded away. These environments also shaped the distribution of mineral resources. The processes that built Columbia concentrated metals and other materials within the crust, preserving them for billions of years.

Much of what is mined today traces back to these ancient events.

At the time, Earth itself was more active. The mantle was hotter. Volcanic activity was more widespread. Tectonic forces were more vigorous. The surface of the planet was not stable in the way we experience it now.

It was in motion.

Yet from that motion emerged something remarkable, a single vast continent.

Imagine standing on that ancient surface. There is no familiar map, no recognizable coastline. A continuous landmass stretches across the horizon.

Mountain chains rise in the distance, newly formed and still changing. Rivers cut through fresh rock, carrying sediment across wide plains. In some regions, volcanic activity reshapes the landscape. In others, erosion is already beginning its slow work.

The air is different. The oceans circulate differently. The planet is the same Earth, but it operates under different conditions.

This world does not last.

Like all supercontinents, Columbia eventually began to break apart. Deep forces within the Earth pulled its pieces away from one another. The crust stretched and fractured. New oceans opened. The connected landmass separated into drifting fragments once again.

The process was as slow as its formation, and just as powerful.

Over time, those fragments reorganized into new supercontinents, including Rodinia and, much later, Pangaea. Each would follow a similar pattern of assembly and breakup, driven by the same forces operating beneath the surface.

Columbia disappeared as a landmass.

Its influence did not.

The mountain ranges that once marked its collisions have been erased, but the deeper structures formed during those events remain embedded within the continents. These structures continue to shape the architecture of the crust and the distribution of mineral resources.

In this way, Columbia survives.

Not as a place you can see, but as a structure you can trace.

The rocks beneath our feet preserve the clues that allow geologists to reconstruct a world that existed more than a billion years ago. They record the formation of continents, the collision of landmasses, and the deep processes that shaped the planet.

At the center of that story are the cratons, the ancient fragments that formed the foundation of the earliest continents.

They appear isolated today.

They are not.

They are pieces of something larger, carried across time and space, still holding the record of when they were joined (see Figure 1.1).

Understanding Columbia begins with them.

Because before there was a supercontinent, there were fragments.

And those fragments still remain.

Figure 1.1 — Reconstruction of Columbia

~1.8 billion years ago

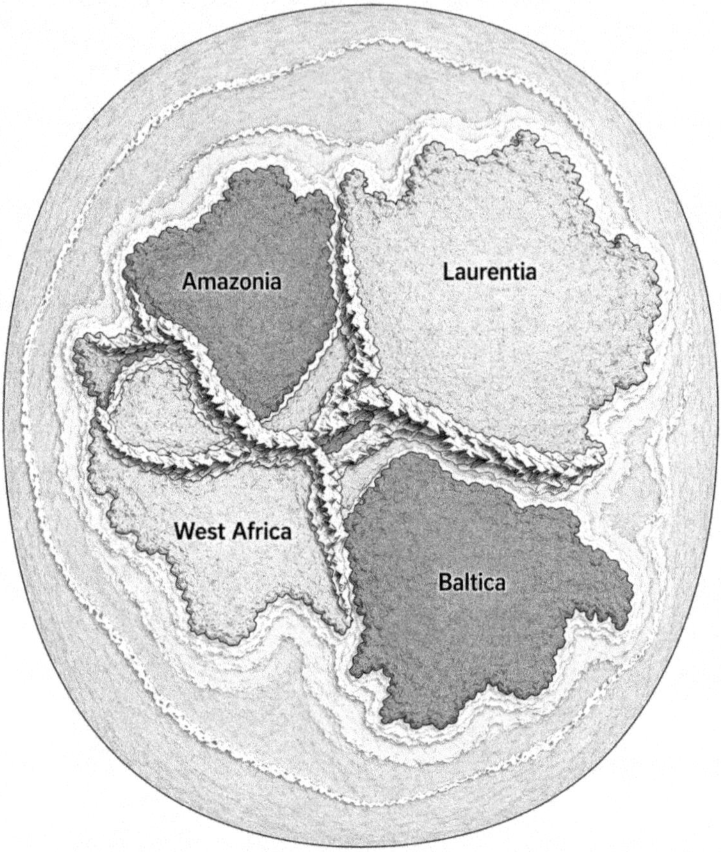

Columbia was assembled by the collision of ancient cratons.
Mountains formed along the sutures; oceans closed.

- Laurentia
- Baltica
- Amazonia
- West Africa
- Orogenic belts (sutured margins)
- Ancient ocean basins

Chapter 2

Fragments of the Ancient Earth

If you were to strip away the forests, the soil, and the cities, and look directly at the oldest parts of the continents, you would see something very different from the landscapes we usually imagine.

In northern Canada, vast areas of exposed rock stretch for hundreds of miles. The surface is uneven but subdued, low ridges, shallow basins, lakes filling depressions carved long ago. The rock is hard, dense, and worn smooth in places by repeated cycles of erosion.

It does not look like the remains of a continent in motion.

It looks permanent.

Yet this is some of the oldest surviving crust on Earth.

These regions are known as shields, places where ancient continental rock is exposed at the surface. Beneath them lie the cores of continents, structures that have endured while much of the rest of Earth's crust has been created and destroyed over and over again.

To understand Columbia, we begin here.

Long before the continents we know today took shape, Earth's surface was not organized into large, stable landmasses. Instead, it was a shifting patchwork of smaller fragments drifting slowly across the planet.

These fragments are known as cratons.

They are the earliest surviving pieces of continental crust.

Most of Earth's early crust did not survive. The planet's surface has always been shaped by motion. New crust forms, old crust is destroyed, and material is continually recycled into the interior.

This is especially true of the ocean floor.

Oceanic crust forms at mid-ocean ridges, where magma rises and solidifies to create new seafloor. But it does not last. As tectonic plates move, oceanic crust is eventually pushed toward regions where one plate sinks beneath another. This process, known as subduction, carries the crust back into the mantle system.

There, it is heated, transformed, and ultimately recycled.

Because of this cycle, the ocean floor we see today is relatively young. Most of it is less than two hundred million years old, a brief span in geological terms.

Continental crust behaves differently.

It is thicker, lighter, and more resistant to being pulled down into the mantle. When continents collide, the crust deforms and thickens, but much of it remains at the surface rather than being recycled.

This difference allows certain fragments of crust to endure.

Over immense spans of time, these fragments became the stable cores of continents.

These are the cratons.

What makes a craton unique is not just its age, but its stability.

Beneath each craton lies a deep, rigid structure that extends far into the mantle. This is known as a mantle keel, a thick root of cool, dense rock that anchors the crust above it. These keels can extend to depths of 150 to 200 miles.

To picture this, imagine a structure far larger below the surface than above it, like an iceberg with most of its mass hidden from view.

The visible crust is only part of the system.

The deeper root is what gives it strength.

Because these roots are cool and rigid compared to the surrounding mantle, they resist deformation. While nearby regions may stretch, fracture, or melt, the craton remains comparatively intact.

It moves with the plate, but it does not easily break.

This stability allows cratons to preserve Earth's geological memory.

The rocks within these regions record processes that occurred billions of years ago, conditions that no longer exist on the planet today. In some places, erosion has stripped away younger layers, exposing this ancient crust at the surface.

That is what we see in shields.

Across the modern continents, these fragments are scattered.

The Canadian Shield in North America is one of the largest and best exposed. Similar regions exist in West Africa, northern Europe, Siberia, South America, and parts of Australia and Asia. Each preserves a piece of Earth's earliest continental foundation.

At first glance, they appear isolated.

Separated by oceans, shaped by different climates, and embedded within entirely different continents, they seem unrelated.

But when geologists study them in detail, patterns begin to emerge.

The ages of rocks match across continents. Structures align when the continents are repositioned. Mineral deposits appear in parallel zones. Even the deep roots beneath these regions share similar characteristics.

These similarities are unlikely to be random.

They point to a time when these fragments were not scattered, but connected.

To see this, it helps to think again in terms of time.

At the speed tectonic plates move, continents drift only a few centimeters each year. It is not a motion we can observe directly. But given enough time, it reshapes the planet.

Over 10 million years, a fragment may move the distance across a small sea.

Over 100 million years, it can cross an ocean basin.

Over a billion years, it can be part of multiple supercontinents, assembled, broken apart, and assembled again.

What appears fixed is not fixed at all.

Cratons are not static remnants.

They are mobile fragments with long histories.

They formed under conditions very different from those of the present Earth. They survived when much of the surrounding crust did not. And over time, they moved, collided, and merged with other fragments.

They are pieces of a larger system.

Columbia was built from them.

Understanding that supercontinent does not begin with a map. It begins with these fragments, the oldest surviving pieces of the Earth itself.

They are the foundation on which everything else was built.

And they carry the record of how it happened.

Chapter 3
A Hotter, Younger Planet

To understand how the earliest continents formed, and how the cratons that built Columbia came into being, you have to imagine a planet that does not behave the way ours does now.

The Earth today feels stable. The ground beneath us shifts too slowly to notice. Mountains rise and erode over spans far longer than human history. The atmosphere is balanced. The oceans are familiar.

But early Earth was not like this.

It was hotter, more active, and far less stable.

And that difference began deep within the planet.

When Earth formed approximately 4.5 billion years ago, it did not emerge as a quiet, solid world. It formed through violent collisions as smaller bodies merged together. Each impact released enormous amounts of energy. Material compressed under its own gravity, generating additional heat. Radioactive elements within the planet began to decay, releasing still more energy into the interior.

The result was a planet filled with heat.

Even after the surface cooled enough for a crust to form, that heat did not disappear. It remained trapped within the mantle and core, slowly moving outward over time.

Early in Earth's history, there was more of it.

Radioactive elements such as uranium, thorium, and potassium were more abundant than they are today. As they decayed, they produced greater amounts of heat, driving stronger motion within the planet's interior.

The engine of the Earth was likely running faster.

Today, the mantle moves slowly. Tectonic plates shift a few inches each year. Over millions of years, that motion rearranges continents, but the pace feels almost static on a human scale.

On the early Earth, that motion was likely more vigorous.

Higher temperatures would have allowed the mantle to flow more easily, increasing the speed and intensity of convection. Plates may have moved more rapidly. Subduction zones may have been more active. The boundaries between plates may have been more dynamic, with more frequent volcanic activity and crustal recycling.

The surface was in near-constant motion.

Crust formed, shifted, and was pulled back into the mantle. Fragments of early continental crust, the cratons, developed within this system, forming in regions where conditions allowed the crust to thicken and stabilize rather than be destroyed.

They were exceptions in a restless world.

While the interior remained hot, the surface had cooled enough for liquid water to exist.

That change altered everything.

Water vapor in the atmosphere condensed and fell as rain, collecting in low areas to form the first oceans. These oceans covered much of the planet, but they were not like the oceans we know today.

They were darker. Rich in dissolved metals, especially iron, they may have had a greenish tint. Their chemistry reflected a world still dominated by volcanic activity and interaction with newly formed crust.

Over time, chemical reactions within these oceans produced layered deposits of iron-rich sediment on the seafloor. These formations, preserved today as banded iron formations, record the interaction between the oceans, the crust, and the atmosphere during this early period.

The oceans were part of a system still taking shape.

The atmosphere was also very different.

Oxygen, which now makes up about twenty-one percent of the air, was almost entirely absent. Instead, the atmosphere was composed largely of carbon dioxide, methane, nitrogen, and water vapor. Volcanic activity continually released these gases, maintaining an environment that would feel unfamiliar, and likely hostile, by modern standards.

There was no breathable air.

Yet even in this environment, life had already begun.

The earliest organisms were microscopic. Bacteria and archaea lived in the oceans, forming layered structures known as stromatolites. Some of these organisms used sunlight to produce energy through photosynthesis, releasing oxygen as a byproduct.

At first, that oxygen did not accumulate in the atmosphere.

It reacted with dissolved iron in the oceans, forming iron-rich minerals that settled to the seafloor. Over long periods of time, as these reactions consumed available iron, oxygen began to build up in the atmosphere.

This transition, known as the Great Oxidation Event, began around 2.4 billion years ago.

It marked a fundamental shift in the planet's chemistry.

The change did not happen quickly. It unfolded over tens of millions of years. But its effects were permanent. The atmosphere was transformed. New chemical processes became possible. The surface environment began to move toward the conditions we recognize today.

Life, though still simple, was reshaping the planet.

All of these systems were connected.

Heat from the interior drove tectonic motion. Tectonic motion reshaped the crust. The crust interacted with the oceans. The oceans influenced the atmosphere. The atmosphere, in turn, was altered by life.

Nothing operated in isolation.

This was the world in which the earliest continental fragments formed.

The cratons that would later assemble into Columbia developed under these conditions, within a hotter mantle, beneath a different atmosphere, and alongside oceans that carried a different chemistry. They formed where crust thickened enough to resist destruction, where deep roots stabilized the surface, and where fragments could persist long enough to become part of something larger.

They were not inevitable.

They were the result of a balance within a dynamic system.

Over time, these fragments survived while surrounding crust disappeared. They moved across the surface, carried by tectonic plates, and began to interact more frequently as the planet evolved.

A hotter Earth made those interactions more likely.

Movement was faster. Collisions were more frequent. The processes that would eventually assemble a supercontinent were already underway.

The conditions were set.

In the next chapter, those drifting fragments begin to converge, tracing the collisions that reshaped the surface of the Earth and brought the first great landmass together.

Chapter 4
The Collisions of Continents

Long before continents meet, the collision has already begun.

Far out beneath the ocean, where no land is visible, the crust between two drifting fragments is slowly disappearing. Year by year, inch by inch, the ocean floor is pulled downward into the mantle along a subduction zone. There is no sudden moment, no single event that marks the beginning.

It is a process unfolding over millions of years.

At the surface, the ocean may appear unchanged. Waves move across it. Sediment settles quietly on the seafloor. But far below, the distance between continents is closing.

Eventually, the ocean itself is consumed.

This is how most continental collisions typically begin, not with land striking land, but with the disappearance of the ocean that separates them.

As one tectonic plate is forced beneath another, the descending slab carries oceanic crust deep into the Earth. Heat and pressure transform it. Water released from the slab alters the surrounding mantle, lowering its melting point and generating magma. That magma rises, forming volcanic arcs that may stretch for hundreds or even thousands of miles along the edge of a continent.

For a long time, this is the dominant process.

The ocean narrows. The distance shrinks. Volcanic chains mark the boundary where one plate descends beneath another. From the surface, the continents still appear separate.

But they are moving toward each other.

When the last of the oceanic crust is gone, the nature of the system changes.

The continents themselves begin to collide.

Unlike oceanic crust, continental crust is too buoyant to sink easily into the mantle. When two continents meet, neither is readily pulled downward. Instead, the crust between them compresses, thickens, and deforms.

The collision does not produce a clean boundary.

It produces a zone of intense transformation.

Layers of rock that once lay flat are folded, tilted, and stacked. Massive sheets of crust are pushed over one another. The surface shortens horizontally, but the crust thickens vertically. Material that was once spread over a wide area is compressed into a smaller space.

The result is the formation of mountain ranges (see Figure 4.1).

To picture this, it helps to think in slow motion.

If two rigid objects collide quickly, they shatter. But if the motion is slow and continuous, the material deforms instead. It bends, folds, and thickens. Over millions of years, the crust behaves less like brittle rock and more like a material that can flow under pressure.

Figure 4.1
Stages Leading to Continental Collision

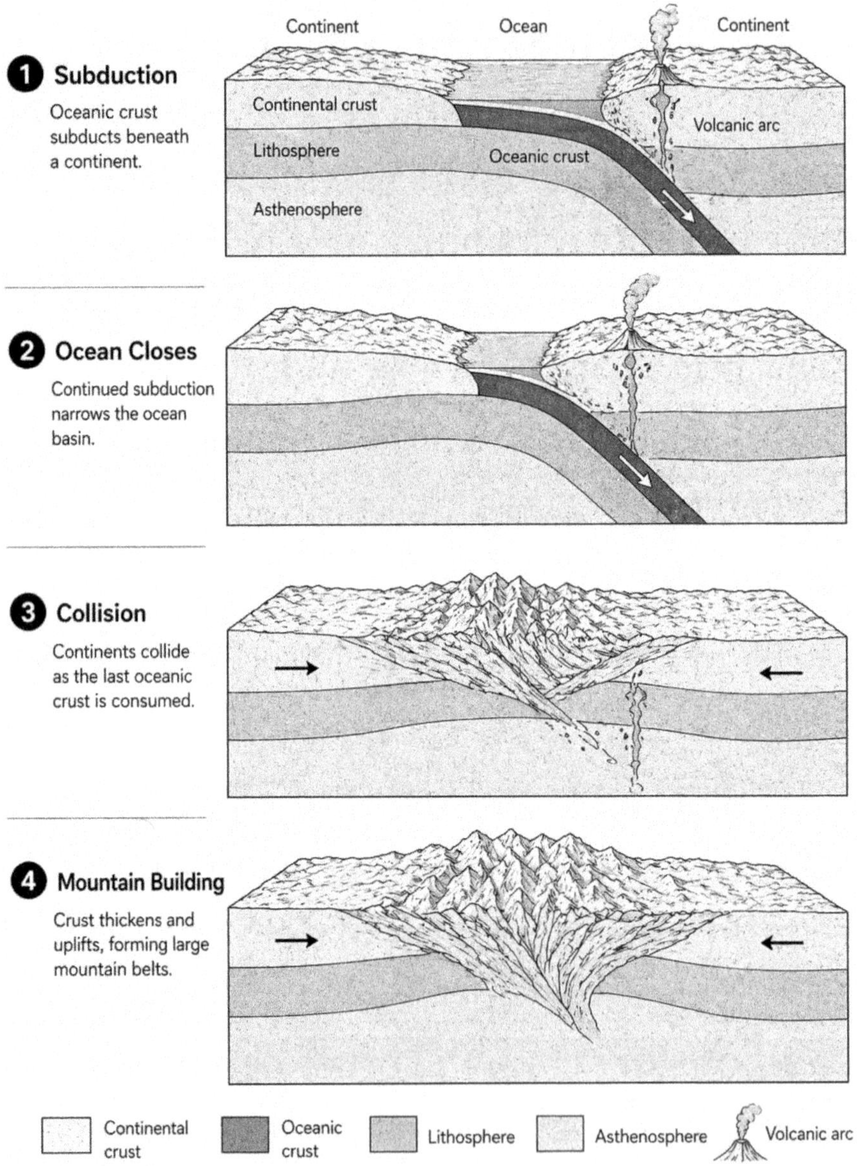

The mountains that form are not sudden features.

They rise gradually, over tens of millions of years, as the collision continues. Peaks grow higher as the crust thickens. Deep valleys form between ridges. Entire regions are uplifted.

A modern example can be seen in the ongoing collision between the Indian and Eurasian plates. India continues to move northward into Asia at a rate of a few centimeters per year. That motion is barely noticeable on a human timescale.

But over millions of years, it has produced the Himalayas, the highest mountain range on Earth.

The Himalayas are not a completed feature.

They are still rising.

This ongoing collision provides a glimpse into processes that shaped the ancient world. During the formation of Columbia, similar events occurred repeatedly as cratons moved across the surface and converged.

Each collision added to the growing landmass.

But the most important changes were not only at the surface.

They occurred deep within the crust.

As the crust thickens, rock is forced downward as well as upward. Material that was once near the surface is buried to great depths, where pressure and temperature increase dramatically. Under these conditions, rocks undergo metamorphism, their structure and composition change without fully melting.

Minerals reorganize. New structures form. The rock records the conditions it experienced.

In some regions, temperatures become high enough for partial melting to occur. This produces magma, which begins to rise through the crust. Some of it reaches the surface as volcanic activity. Much of it remains below, cooling slowly to form large bodies of igneous rock.

At the same time, fluids are released within the crust.

Water and other volatile components, trapped within minerals, are driven out by heat and pressure. These fluids move through fractures created by the collision, dissolving elements from the surrounding rock and transporting them over long distances.

As conditions change, temperature drops, pressure shifts, or chemistry evolves, those elements are redeposited.

This is one way mineral deposits begin to form.

These processes are not separate.

They are connected.

Crustal thickening increases heat and pressure. Heat drives melting. Melting produces magma. Pressure and temperature release fluids. Fluids transport elements. Together, they reshape the crust and concentrate materials within it.

The surface expression, the mountains, is only one part of the system.

Beneath those mountains, the structure of the crust is being permanently altered.

Even after the mountains erode, those deeper changes remain.

The boundaries where continents join are preserved as suture zones, the seams of ancient collisions. These zones mark the location of former ocean basins that were consumed before the continents met.

Within them, the rock is intensely deformed.

Layers are folded and fractured. Fragments of oceanic crust may be trapped between the colliding continents, preserving pieces of the vanished seafloor. These remnants provide direct evidence of oceans that no longer exist.

Suture zones are complex.

They are also essential.

They record where continents came together. They preserve the history of collision. And they often mark regions where mineral deposits are concentrated, making them important not only for understanding the past, but for locating resources in the present.

During the assembly of Columbia, these collisions appear to have occurred repeatedly.

Each event brought separate fragments together. Each added to the growing landmass. Each thickened the crust, created mountain belts, and left behind structures that would persist long after the surface changed.

The process was slow, but it was cumulative.

What began as scattered fragments became a connected system.

The mountains that formed during these collisions have long since disappeared. Erosion has worn them down. Time has removed their visible form.

But the deeper record remains.

The seams of those collisions still run through the continents. The structures formed beneath ancient mountain ranges are still embedded within the crust. The materials concentrated during those events are still present.

The surface has changed.

The deeper structure remains.

In the next chapter, we turn to those vanished mountains, tracing how something so large could disappear, and what remains after it is gone.

Chapter 5

Mountains That No Longer Exist

If you stand in a place like the Canadian Shield today, you won't see mountains.

You will see open rock, low ridges, scattered lakes, and wide stretches of land that feel quiet and stable. The terrain is subdued. Nothing rises sharply. Nothing suggests scale.

There are no towering peaks. No jagged skylines.

Nothing that signals what once stood here.

And yet, this is where some of the largest mountain systems on Earth once existed.

During the assembly of Columbia, continental collisions created vast mountain ranges that extended across large portions of the supercontinent. These were not isolated peaks. They were continuous systems, stretching for thousands of miles, rising over long periods as the crust thickened and deformed.

They shaped the surface.

They influenced climate, redirected rivers, and controlled how sediment moved across the land. They marked the boundaries between colliding fragments and recorded the forces that brought those fragments together.

But if you look for them now, they cannot be found.

Their disappearance is not unusual.

Mountains do not persist.

From the moment they begin to rise, they are also being worn down. The same forces that build them are balanced by processes that reduce them, slowly at first, then steadily over time.

Water is one of the most effective.

Rain falls onto exposed rock and works its way into small fractures. Over time, those fractures widen. Pieces break free. Rivers carry that material downhill, transporting sediment across valleys and eventually toward the oceans.

In colder regions, ice adds another force.

Glaciers move slowly across the landscape, grinding the rock beneath them. They carve valleys, smooth surfaces, and break larger formations into smaller fragments. When the ice retreats, it leaves behind a reshaped terrain and large volumes of debris.

Wind contributes as well, particularly in drier environments.

It lifts fine particles and carries them away, gradually stripping exposed surfaces. Its effects are slower, but over long periods, they accumulate.

These processes do not act once.

They act continuously.

At first, the changes are small. A ridge lowers slightly. A valley widens. A peak loses a small amount of height. Nothing appears dramatic on a human timescale.

But over millions of years, the change becomes substantial.

Mountain ranges are reduced to hills.

Over longer spans, even those hills are worn down, leaving broad plains or gently undulating terrain. What once stood at great height is lowered close to sea level.

The mountain belts that formed during the assembly of Columbia have been subjected to this process for more than a billion years.

Given enough time, removal is not partial.

It is complete.

But the mountains did not only extend upward.

They extended downward.

When continents collide, the crust thickens in both directions. Material is pushed upward to form mountains, but it is also forced deeper into the Earth, creating a root beneath the range that is far larger than what is visible at the surface.

The higher the mountains, the deeper the root (see Figure 5.1).

This deeper structure is not exposed when the mountains are first formed. It lies far below, under conditions of high pressure and temperature, where the rock behaves differently than it does at the surface.

As erosion removes the mountains above, that deeper material does not disappear.

It becomes exposed.

Rock that once existed miles beneath the surface is gradually brought upward, not by movement, but by removal of the material above it. Over time, the upper layers are stripped away, revealing the

Figure 5.1
Deep roots of ancient mountain belts
Mountains form during continental collision

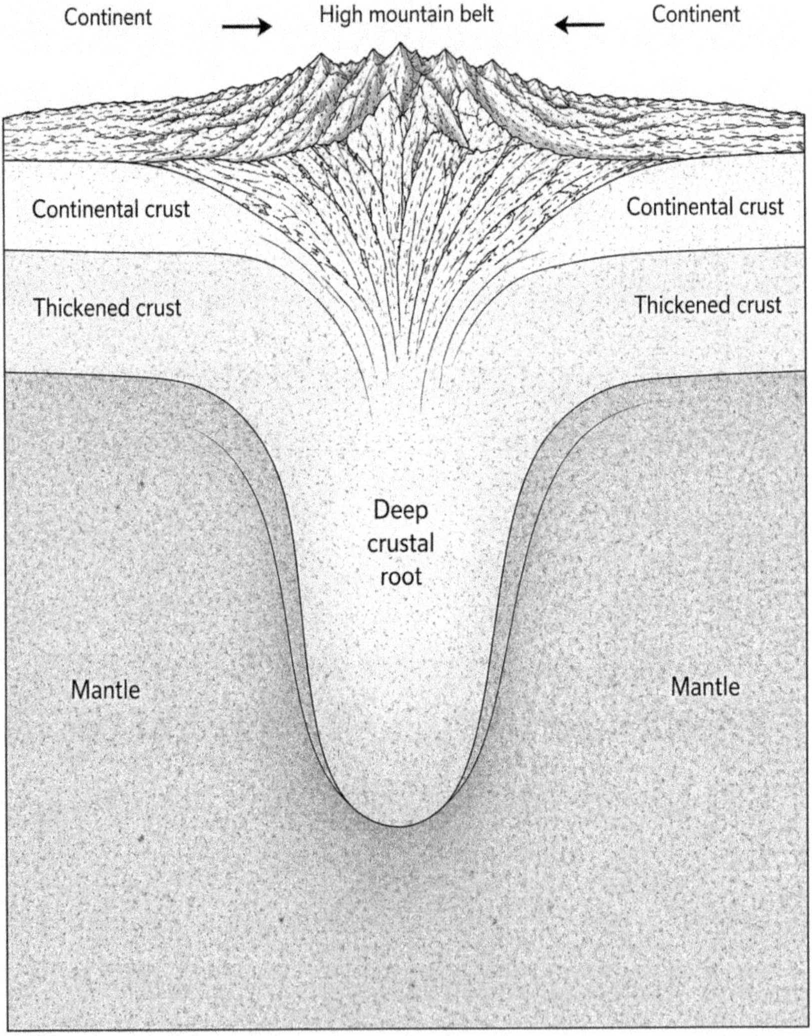

internal structure of the former mountain system.

This is what remains.

Metamorphic rock.

Under high pressure and temperature, existing minerals reorganize. New minerals form. The rock records the conditions it experienced deep within the crust. Its structure reflects alignment under stress. Its composition reflects the environment in which it was altered.

These rocks appear today as elongated zones across the continents.

They are not random.

They are the preserved cores of ancient mountain belts.

Geologists refer to them as metamorphic belts, regions where rock has been deeply altered and deformed during continental collision. They often extend across large distances, marking where fragments once joined.

At the surface, they do not resemble mountains.

They appear as exposed rock, often low and eroded, with little to indicate their origin.

But when studied, their structure reveals the past.

Layers are folded. Minerals are aligned. Boundaries mark zones of compression and deformation. These features trace the internal architecture of mountain systems that no longer exist.

Across modern continents, these belts appear separated.

But when those continents are repositioned, they align.

What seems fragmented becomes continuous.

A belt in one region can connect with another across an ocean. The structure matches. The sequence of deformation is consistent. The rocks record the same history.

The mountains are gone.

The structure remains.

This structure does more than record collision.

It also preserves the conditions that concentrated elements within the crust. Heat, pressure, and fluid movement during mountain building created environments where metals could accumulate. As fluids moved through fractures, they dissolved elements and redeposited them in concentrated zones.

These processes occurred within the same systems that formed the mountains.

When the mountains were removed, the deposits remained.

Many of the resources used today, gold, copper, and other metals, are linked to these ancient environments. They are not distributed randomly. They follow the structures formed during collision and preserved within the crust.

The visible landscape offers little indication of this history.

A region may appear flat or gently rolling, with no sign that it once held large mountain systems.

But the record is still present.

It exists in the rock, in its structure, its composition, and its arrangement across the continents.

The mountains of Columbia have been erased from the surface.

Their roots have not.

And it is within those roots that the next part of the story continues, deeper within the Earth, where the forces that built and reshaped the continents are still at work.

Chapter 6
Fire Beneath the Crust

The surface of the Earth can feel still.

Mountains appear fixed. Continents seem anchored in place. Landscapes change so slowly that, within a human lifetime, they feel permanent.

But that stability is an illusion.

Beneath the crust, the planet is in motion.

Deep within the Earth, heat moves through the mantle in vast, slow currents. The mantle is solid rock, but under the immense pressures and temperatures at depth, it behaves differently than rock at the surface. Over long periods of time, it can flow.

Not quickly. Not visibly.

But continuously.

This movement carries heat from the deep interior toward the surface. It is not uniform. Some regions rise as hotter material moves upward. Others sink as cooler material descends. Together, these motions form a system of convection, an ongoing circulation driven by differences in temperature and density.

This is the primary engine of the Earth.

It operates at a pace that is difficult to grasp. In a single year, the movement may be measured in inches. Over thousands of years, it remains subtle. But over millions of years, it reshapes the entire surface of the planet.

On the early Earth, that engine ran hotter.

Greater internal heat meant stronger convection. Material within the mantle moved more vigorously. The forces transmitted to the surface were more intense, driving faster tectonic motion and more active geological processes.

The crust above responded to this motion.

Tectonic plates, the large segments of Earth's outer shell, are carried along by the movement of the mantle beneath them. As they move, they interact in different ways, depending on their direction and the forces acting on them.

In some regions, plates move apart.

As the crust is pulled in opposite directions, it begins to thin. The pressure on the underlying mantle decreases, allowing hot material to rise toward the surface. As it rises, part of it melts, forming magma.

This magma fills the gap.

It cools and solidifies, creating new crust.

Over time, this process forms long ridges beneath the oceans, where new seafloor is continuously generated. What begins as a narrow fracture can expand into a wide ocean basin.

In other regions, plates move toward each other.

When this happens, one plate may be forced beneath another in a process known as subduction. As the descending plate sinks into the mantle, it carries water and other materials with it. These components are released under increasing pressure and temperature, lowering the melting point of the surrounding rock.

Magma forms again, but through a different pathway.

This magma rises through the crust, feeding volcanic systems that often form long chains known as volcanic arcs. These arcs can extend for hundreds or even thousands of miles, marking the boundary where one plate descends beneath another.

These two settings, where plates pull apart and where they collide, account for much of the planet's volcanic activity.

But they are not the only ones.

In some cases, heat rises from deep within the mantle in focused columns known as plumes. These plumes originate far below the surface and carry hot material upward over long periods of time.

When a tectonic plate moves across a relatively stationary plume, a chain of volcanoes can form.

The Hawaiian Islands provide a modern example. Each island represents a position where the plate passed over the plume, leaving a trail that records movement over millions of years.

On the early Earth, when internal heat was greater, these plumes may have been more common and more intense.

They added another layer to an already dynamic system.

Although these processes appear different at the surface, they are connected by the same underlying cause.

Heat.

Heat drives convection. Convection moves plates. Plate movement creates regions of separation, collision, and upwelling. Each setting produces magma through slightly different mechanisms, but all are part of the same system.

This system does more than move rock.

It redistributes it.

As magma forms and rises through the crust, it carries dissolved materials with it, elements such as copper, nickel, and gold. As the magma cools, these elements can become concentrated, separating into different minerals at different stages of crystallization.

Some remain locked within solid rock.

Others are carried further.

At the same time, fluids move through the crust.

Water and other volatile components, released by heat and pressure, travel along fractures and faults. As they move, they dissolve elements from surrounding rock. When conditions change, those elements are redeposited, forming concentrated mineral deposits.

These processes help link the deep Earth to the surface.

Material is not static.

It moves, transforms, and is redistributed across the crust over long periods of time.

Carbon follows similar pathways.

Deep within the mantle, under extreme pressure, carbon can crystallize into diamond within the stable roots of cratons. In other environments, it returns to the surface through volcanic systems, released as carbon dioxide.

The same system that builds continents also cycles elements through the planet.

Nothing is isolated.

Mantle convection drives tectonic motion. Tectonic motion generates magma. Magma alters the crust and transports elements. Fluids redistribute those elements further. Over time, these processes build the structure of continents, form mountain belts, and create the conditions that preserve Earth's geological history.

The surface may appear stable.

It is held in place by motion below.

This motion was especially important during the early history of the Earth. A hotter interior drove stronger processes, increasing the likelihood that crust would form, interact, and evolve into stable structures.

It created the conditions in which cratons could develop.

It powered the collisions that assembled Columbia.

And it shaped the deep structures that still influence the planet today.

Among the most remarkable products of this system are diamonds, crystals formed far beneath the surface, within the deep roots of ancient continents, and preserved for billions of years.

In the next chapter, we follow those diamonds, tracing their origin deep within the mantle and their journey to the surface.

Chapter 7
Diamonds from the Deep

Most minerals form close to the surface.

They crystallize from cooling magma, settle from water, or grow within rocks under conditions that exist within the crust. Their origins are tied to environments we can observe, study, and, in many cases, visit directly.

Diamonds are different.

They form far below the reach of direct observation, deep within the Earth, under conditions that exist only within a narrow range of pressure and temperature. They are not products of the surface environment. They are fragments of the deep Earth.

To understand where diamonds come from, it helps to return to the cratons, the oldest and most stable parts of the continents. Beneath each craton lies a thick, rigid root of mantle rock extending deep into the Earth. These structures, known as mantle keels, create a unique environment.

They are cool relative to the surrounding mantle.

They are stable over immense spans of time.

And they extend to depths where pressure becomes extreme.

Within these roots, conditions can reach the range required for diamond formation. Carbon, present in the mantle, is subjected to pressures and temperatures that force its atoms into a tightly bonded crystal structure.

The result is diamond.

This structure is what gives diamond its remarkable properties. It is composed entirely of carbon, just like graphite. But the arrangement of atoms is different. In graphite, carbon forms layers that can slide over one another. In diamond, the atoms are locked into a rigid, three-dimensional lattice.

The difference is structural.

Under the right conditions, carbon becomes one of the hardest natural materials known.

Those conditions are rare.

They exist only within specific regions of the mantle, typically beneath ancient cratons. Without the stability provided by these deep roots, diamonds would not form, or, if they did, they would not survive.

Many diamonds formed more than two and a half billion years ago, during the early history of the Earth. They developed deep within the mantle and remained there, locked within the roots of continents as those continents moved, collided, and reassembled over time.

A single diamond may have existed through multiple supercontinent cycles.

It may have been present during the assembly of Columbia, persisted through its breakup, and remained deep within the Earth as new continents formed and separated again.

For most of its existence, a diamond remains hidden.

It does not move on its own. It does not rise slowly through the crust. It remains where it formed, stable under the conditions that created it.

Reaching the surface requires something rare.

A rapid, violent event.

Diamonds are transported upward by deep-source volcanic eruptions known as kimberlite and lamproite eruptions. These eruptions originate deep within the mantle and rise quickly through the crust, carrying fragments of mantle material with them.

Speed is critical.

If the ascent is too slow, the diamond will not survive the journey. As pressure decreases, the stable form of carbon changes. Diamond will revert to graphite under lower-pressure conditions unless it is carried rapidly to the surface.

Kimberlite eruptions move fast enough to preserve it (see Figure 7.1).

They travel through narrow conduits, forming structures known as kimberlite pipes. These pipes cut through the crust, creating vertical pathways that bring deep material to the surface in a relatively short period of time.

When the eruption ends, the pipe remains.

Over millions of years, erosion removes surrounding rock, exposing these structures. Diamonds contained within them may be released into the surrounding environment, carried by rivers, and concentrated in sedimentary deposits.

Figure 7.1 — Diamond formation

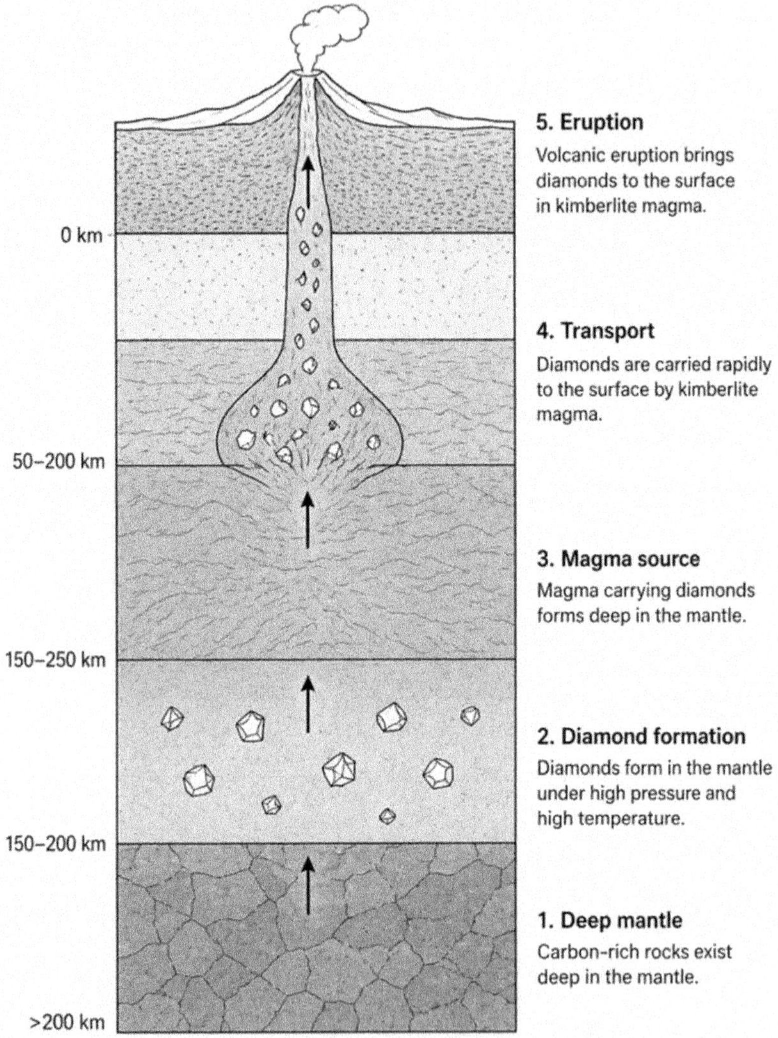

Diamonds form deep in the mantle and are delivered to the surface by volcanic eruptions.

This is how many diamonds are eventually discovered, not at their source, but downstream, in alluvial deposits formed long after the original eruption.

Because diamond formation depends on the deep roots of cratons, most major deposits occur in regions underlain by ancient continental crust.

Southern Africa, Canada, Russia, and parts of Australia all sit atop these old geological foundations. Though widely separated today, these regions share a common origin in the early development of continents.

They are tied to the same deep structures.

Each diamond carries a record of that history.

Some contain microscopic inclusions, tiny fragments of surrounding material trapped within the crystal as it formed. These inclusions preserve information about pressure, temperature, and chemical conditions deep within the mantle.

They are samples of a place that cannot be directly reached.

Through them, scientists can study environments hundreds of miles below the surface, reconstructing conditions that would otherwise remain unknown.

A small number of diamonds originate even deeper.

These "superdeep" diamonds form at depths far below typical cratonic roots, suggesting that carbon is cycled through the mantle over a much greater range than once thought. They provide evidence that processes within the Earth may extend deeper and operate more broadly than earlier models suggested.

Even within the category of diamonds, there is variation.

But the common thread remains the same.

They are products of the deep Earth, preserved within stable regions, and brought to the surface by rare events.

Their journey is long, spanning nearly the entire geological history of the planet

From formation deep within the mantle, to billions of years of stability, to rapid transport through volcanic systems, to eventual exposure at the surface, a diamond's history spans nearly the entire geological evolution of the planet.

What appears small and isolated is part of something much larger.

Diamonds are not just gemstones.

They are evidence.

They show that the deepest parts of the Earth remain connected to the surface through dynamic processes. They demonstrate the importance of cratons as stable environments capable of preserving material over immense spans of time. And they provide direct insight into conditions that cannot otherwise be observed.

They are one of the clearest examples of how the oldest parts of the Earth continue to influence the present.

The same deep systems that formed diamonds also concentrate metals within the crust, materials that shape industries, technologies, and economies.

In the next chapter, we follow those metals, tracing how ancient collisions created the conditions for their formation and how those processes continue to shape the distribution of resources today.

Chapter 8
Metals of Ancient Collisions

When continents collide, the most visible result is the landscape.

Mountains rise. Valleys form. The surface reshapes across vast regions. But the most important changes are not always visible.

They occur within the crust.

As collision thickens the crust, conditions change. Temperature increases. Pressure rises. Rock deforms, fractures, and, in some regions, begins to melt. Within this environment, elements that were once dispersed begin to move.

When they move, they begin to concentrate.

Over time, this concentration creates mineral deposits, zones where specific elements accumulate in far greater amounts than in the surrounding rock. These zones can remain stable for billions of years, preserved long after the mountains that formed them have been removed.

The process follows the same general pattern across different regions.

But the details vary.

Different elements concentrate under different conditions, depending on how heat, magma, and fluids interact within the crust.

Nickel and iron are commonly associated with deep magmatic systems.

In some regions, large volumes of magma form at depth and begin to cool slowly. As they cool, dense minerals separate from the melt and settle downward. These minerals carry elements like nickel and iron with them, forming layered deposits within the crust.

These systems develop over long periods.

They are tied to large, stable environments where magma can accumulate and evolve rather than erupt quickly to the surface.

Copper forms under different conditions.

It is commonly linked to volcanic arc environments, where magma rises through the crust above subduction zones. As that magma moves upward, it releases fluids rich in dissolved elements. These fluids circulate through surrounding rock, moving along fractures and faults.

As conditions shift, temperature drops, pressure changes, or chemical balance is altered, the copper is deposited (see Figure 8.1).

Not uniformly, but in concentrated zones that can extend across large areas.

Gold often follows yet another path.

It is often associated with the same regions that record intense deformation during continental collision, metamorphic belts formed under high pressure and temperature. In these environments, fluids move through networks of fractures created as the crust is compressed.

Figure 8.1
Hydrothermal circulation

Seawater infiltrates the crust, is heated by magma, rises through fractures, precipitates minerals near the seafloor, and returns—driven by heat.

Example deposit: copper (Cu)

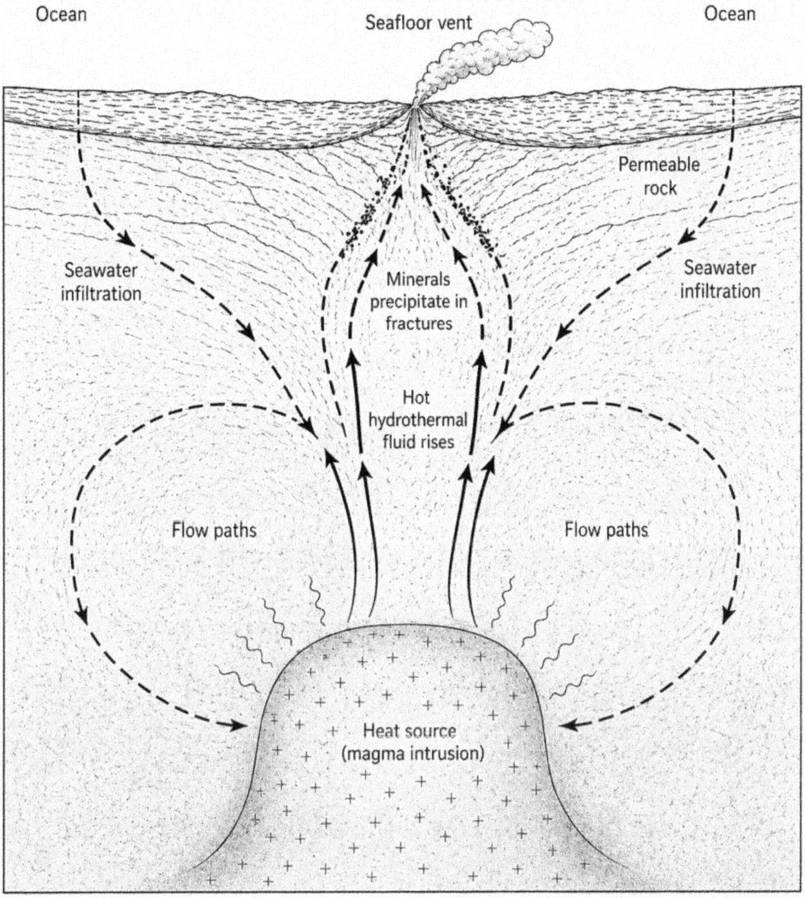

These fluids carry small amounts of gold.

As they move and conditions change, the gold is redeposited, forming veins that can become highly concentrated over time.

These systems are often narrow, but they can extend for great distances, tracing the same structures that record the collision itself.

Iron has a more complex history.

Many major iron deposits began in ancient oceans, where dissolved iron settled to the seafloor in layered formations. During later continental collisions, these formations were uplifted, deformed, and incorporated into the continental crust.

What formed in one environment was reshaped in another.

Uranium tends to concentrate in ancient continental regions where fluid movement and chemical conditions allow it to accumulate within sedimentary basins and fracture systems. These deposits often form later, but they still follow the structural framework created by earlier events.

Each of these elements follows a different pathway.

But those pathways are not independent.

They occur within the same evolving system.

Collision thickens the crust. Heat increases. Magma forms in some regions. Fluids move through others. Fractures provide pathways. Changes in temperature, pressure, and chemistry determine where elements are carried and where they are left behind.

The result is not random in distribution.

It is patterned.

Across the assembly of Columbia, these processes operated across large regions, creating mineral belts that extended across the connected landmass. When the supercontinent later broke apart, those belts did not disappear.

They were carried with the fragments.

Today, they appear scattered across continents that are separated by oceans. A gold-rich region in one part of the world may share its origin with a similar region thousands of miles away. A sequence of copper deposits may form a continuous arc when those continents are repositioned.

What appears isolated is part of a larger system.

These patterns are not only records of the past.

They shape the present.

Many of the materials used in modern society, metals essential for construction, technology, and energy, are drawn from deposits formed during these ancient processes. Their distribution is not accidental. It reflects the deep history of continental assembly and collision.

In this way, the legacy of Columbia remains embedded within the crust.

The mountains that formed during its assembly are gone. The landmass itself no longer exists. But the processes that shaped it left behind structures and materials that persist.

They can be traced.

They can be mapped.

And through them, the structure of a world that no longer exists begins to reappear.

Chapter 9
Mapping the Fragments

By this point, the surface story is gone.

The mountains that marked the collisions have been worn down. The coastlines that defined the edges of Columbia no longer exist. The landmass itself has been broken apart and carried across the globe.

What remains is not a landscape.

What remains resolves into a pattern.

At first glance, that pattern is difficult to see. The continents are separated by vast oceans. Each has its own geography, climate, and geological history layered over billions of years. There is no obvious map connecting them.

But the connection is there.

It is embedded within the crust.

To reconstruct Columbia, geologists do not begin with coastlines. Coastlines change too easily. They are shaped by erosion, sea level, and recent tectonic activity. Instead, geologists look deeper, into the structure, composition, and age of the rocks themselves.

They look for continuity where none appears to exist.

The process is like assembling a puzzle without edges.

There is no guiding frame. Many pieces are missing. Others have been altered or worn down. The image is incomplete, and yet, patterns begin to emerge.

The first clues come from rock ages.

Across continents, geologists have identified regions where rocks formed at the same time, often more than a billion years ago. These ages are determined through radiometric dating, a method that measures the decay of radioactive elements within minerals.

When rocks in distant regions share the same age, it suggests they formed during the same geological event.

On its own, that is not enough.

But it is a starting point.

The next layer of evidence comes from structure.

Ancient mountain belts, now reduced to metamorphic zones, preserve the direction and intensity of past deformation. Layers of rock show how they were folded, compressed, and altered under pressure.

When continents are repositioned, these structures often align.

A belt of deformed rock in one region connects with a similar belt on another continent. The direction of compression matches. The sequence of deformation is consistent. What appears separate becomes continuous.

It is as if a seam has been cut and spread apart.

Mineral deposits add another layer.

As seen in earlier chapters, the processes that concentrate metals during continental collision create distinct patterns within the crust. These patterns are not random. They follow the zones where heat, pressure, and fluid movement were most intense.

When these mineral belts are mapped across continents, they often appear as isolated regions.

But when those continents are brought back together, the belts align.

A gold-rich region in one part of the world may connect with another thousands of miles away. A sequence of copper deposits may form a continuous arc when reconstructed. These alignments support the idea that the regions were once part of the same system.

Each line of evidence adds weight.

But the reconstruction does not rely on a single type of data.

Magnetic signatures provide another set of clues.

When molten rock cools, minerals within it align with Earth's magnetic field. Because the orientation of that field has changed over time, these alignments preserve a record of the position of the rock when it formed.

By comparing magnetic data from different regions, geologists can estimate how continents were oriented relative to one another in the past.

This does not produce a precise map.

But it constrains the possibilities.

Seismic and geophysical data add depth to the picture.

By studying how waves move through the Earth, scientists can identify variations in density and structure within the crust. These variations reveal buried features, thickened crust, deep roots, and boundaries that correspond to ancient collisions.

These features are not visible at the surface.

But they are still present.

Taken together, these lines of evidence form a network of constraints.

No single piece is definitive. But when multiple lines of evidence point to the same configuration, the reconstruction becomes increasingly reliable.

To see how this works, imagine focusing on a single region.

A geologist begins with a craton, an area of ancient crust exposed at the surface. Samples are collected. Minerals are dated. The age of the rock is established. The structure is mapped, revealing patterns of deformation. Mineral deposits are identified and placed within the broader context of the region.

Then the search expands.

Similar ages are found on another continent. Structures appear to match when orientations are adjusted. Mineral belts align. Magnetic data suggests compatible positions.

One connection becomes possible.

Then another.

Gradually, a larger structure emerges.

The process is iterative, refining itself with each step.

Continents are repositioned. Evidence is tested. Some configurations fit better than others. As more data is gathered, the model is refined. Connections that once seemed uncertain become more secure. Others are revised or replaced.

The reconstruction evolves.

What begins as scattered fragments becomes a coherent system.

This system is not perfect.

Some pieces are missing. Others have been altered beyond recognition. Erosion has removed vast amounts of material. Later tectonic events have overprinted earlier structures. The record is incomplete.

But it is not erased.

Enough remains to trace the outlines.

When the fragments are brought together, patterns that once seemed isolated begin to connect. Mountain belts extend across continents. Mineral systems form continuous zones. Cratons fit into positions that reflect shared histories (see Figure 9.1).

Columbia begins to take shape.

Not as a precise map with defined coastlines, but as a structure, a configuration of fragments connected through time.

Figure 9.1
Cratons and mineral belts

Cratons are ancient, stable cores of continents. Matching belts across oceans show these continents were once joined.

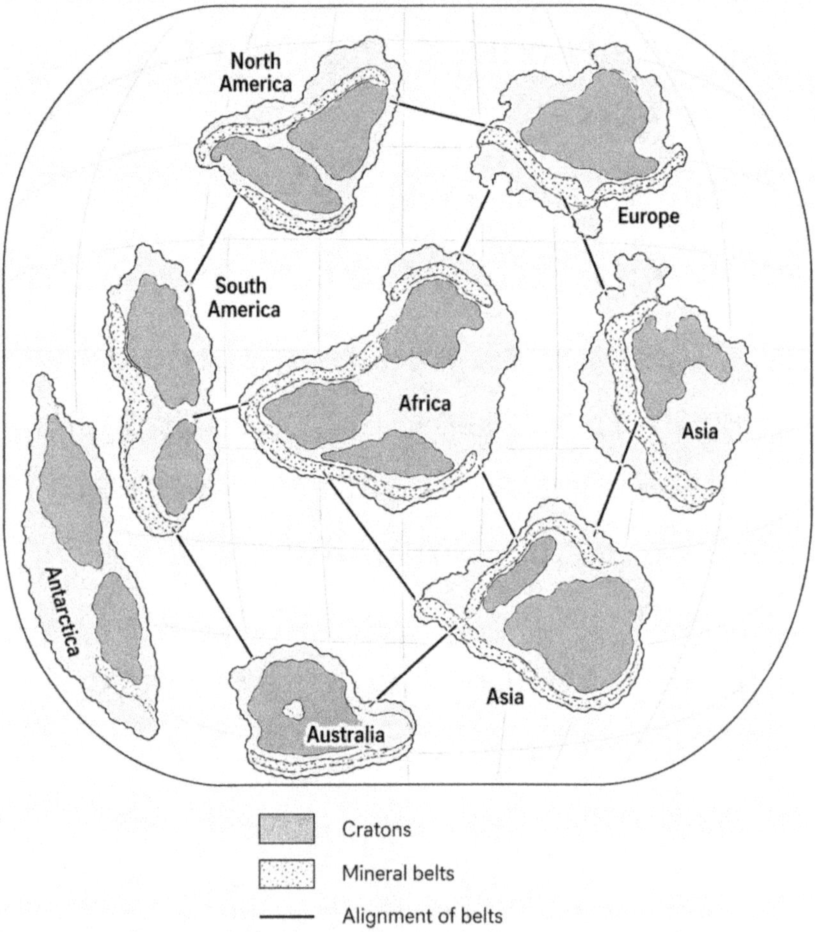

Matching mineral belts across continents reveal ancient connections.

In this sense, Columbia is not something that can be seen directly.

It is something that must be reconstructed.

The continents we see today are the result of billions of years of movement. They carry the record of earlier configurations, but that record is distributed, fragmented, and often hidden.

To read it requires comparison.

It requires assembling pieces that are no longer physically connected.

And it requires accepting that the final picture will never be complete.

Even so, the pattern is clear.

The continents are not isolated.

They are fragments of earlier worlds.

And through the structures preserved within them, those worlds can be traced.

In the next chapter, we follow the process further, examining how the supercontinent itself began to break apart, and how the forces that once brought continents together reversed direction and pulled them apart again.

Chapter 10
The Breakup of Columbia

No supercontinent remains intact.

The same forces that bring continents together eventually begin to pull them apart. The transition is not sudden. It does not begin with a dramatic split or a single defining event.

It begins beneath the surface.

Deep beneath the surface, the system that once supported a unified landmass begins to change.

During the assembly of Columbia, repeated collisions thickened the crust.

Layers of rock were compressed, stacked, and pushed both upward and downward. Over time, this created a vast region of thick continental crust, strong, stable, and continuous across large distances.

At first, this stability holds.

The supercontinent persists. Movement slows. The boundaries that once drove collision become less active. For a time, the system appears balanced.

But beneath that stability, heat continues to rise.

The mantle does not stop moving simply because continents have assembled. Convection continues to carry heat toward the base of the crust. And where the crust is thick, that heat has more difficulty escaping.

It begins to accumulate.

The thickened crust acts like a thermal blanket.

Heat is thought to build beneath it over tens of millions of years. The temperature at depth increases. Rock becomes weaker. The base of the crust begins to soften, losing some of the strength that held the system together.

The change is gradual, unfolding over long periods.

At first, it produces only subtle effects. Small zones of weakness form within the interior of the continent. These may follow older structures, ancient suture zones where continents once collided, or develop in regions where the crust is already under stress.

The system is no longer purely stable, it is becoming unstable from within.

As heat continues to build, the crust begins to stretch.

This stretching does not happen all at once. It begins in localized regions where the crust is weakest. The surface may show long, narrow depressions forming, early rift zones where the land is slowly being pulled apart.

At depth, the changes are more significant.

As the crust thins, pressure on the underlying mantle decreases. This allows hot mantle material to rise toward the surface. As it rises, part of it melts, forming magma that moves into the stretching crust.

This magma fills fractures.

It cools and solidifies, creating new material even as the existing crust is being pulled apart.

The process feeds itself.

Stretching allows mantle material to rise. Rising material produces magma. Magma weakens the crust further, allowing more stretching. Over time, the rift deepens and widens.

What begins as a narrow zone of weakness becomes a major structural feature.

If the process continues, the crust eventually splits.

A new boundary forms.

At that point, the system shifts from continental rifting to seafloor spreading. Magma rises continuously along the boundary, creating new oceanic crust as the two sides move apart.

A new ocean basin begins to open (see Figure 10.1).

Figure 10.1
Continental rifting

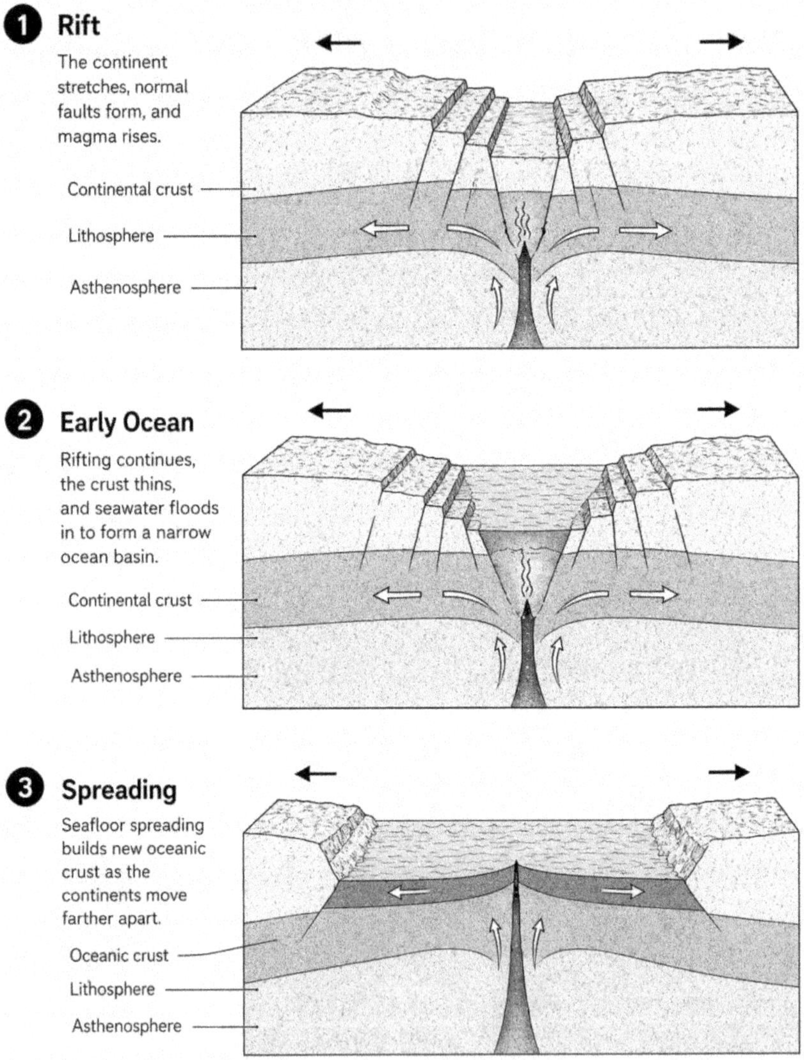

This marks the point at which a supercontinent begins to break apart.

But it does not happen everywhere at once.

The breakup of Columbia unfolded over hundreds of millions of years. Different regions separated at different times. Some fragments remained connected long after others had drifted away. The process was uneven, with multiple rift systems developing and evolving independently.

The result was not a clean division.

It was a gradual fragmentation.

As the pieces moved apart, new coastlines formed. Regions that had once been deep within the interior of the continent became edges. Oceans expanded between the fragments, changing patterns of circulation and, over time, influencing climate and the chemistry of the oceans.

The surface of the planet was being reorganized.

Yet the forces driving this change were the same forces that had built the supercontinent.

Mantle convection continued to move heat through the Earth's interior. Tectonic plates continued to shift. The difference was direction. Instead of convergence and collision, the dominant motion became extension and separation.

The direction of the system had reversed.

What had once pulled fragments together was now pushing them apart.

Even as the surface changed, the deeper structure endured.

The cratons that had assembled to form Columbia did not disappear. Their deep roots remained intact, preserving the stability that had allowed them to survive earlier cycles. As the continent broke apart, these fragments were carried along by the moving plates.

They did not lose their identity.

They were redistributed.

The mountain ranges that once connected them eroded away, but the structures formed during those collisions remained embedded within each fragment. Suture zones, metamorphic belts, and mineral systems persisted, even as the continents separated.

The visible connection was lost.

The structural connection remained.

Over time, the fragments of Columbia continued to move across the planet. They did not drift randomly. Driven by the same underlying forces, they followed paths that would eventually bring them into new configurations.

Hundreds of millions of years later, many of these fragments became part of another supercontinent: Rodinia.

Still later, the process repeated again with the formation of Pangaea.

This repeating pattern, assembly, stability, breakup, is known as the supercontinent cycle.

Columbia was one of its earliest large-scale expressions.

Its breakup did not erase its history.

It scattered it.

The fragments that once formed a single landmass now lie within modern continents, separated by oceans but still carrying the record of their shared past.

The structures created during its assembly continue to influence the arrangement of continents and the distribution of resources.

What appears as separation is part of a longer continuity.

The continents we see today are not final forms.

They are temporary configurations within an ongoing system.

In the next chapter, we begin to trace those fragments in the present, following the clues hidden within today's continents that allow geologists to reconstruct the shape and structure of this lost supercontinent.

Chapter 11

Clues Hidden in Today's Continents

At the surface, the continents appear separate.

They are divided by oceans, shaped by different climates, and defined by coastlines that seem permanent. There is nothing obvious that suggests they were once part of a single landmass.

No visible seams. No clear edges.

The connection is not something you see at a glance.

It is something that must be uncovered.

A geologist does not begin with a map of the ancient world. There is no complete picture waiting to be revealed. The work begins in fragments, often in places that appear ordinary, exposed rock along a roadside, a remote stretch of shield, a quarry face cut into ancient crust.

At first, it looks uniform.

Then something subtle shifts.

A layer tilts slightly in one direction. Minerals align in faint bands. A fracture cuts across the surface at an angle that does not match the surrounding rock. The change is subtle, easy to miss without attention.

So it is measured.

A compass is placed against the rock. The orientation is recorded. Another measurement is taken a few meters away, then another. Individually, each observation is small. Together, they begin to form a pattern.

A fold emerges.

Not as a single feature, but as part of a larger structure extending beyond what can be seen from one position. The rock records movement, compression, a force applied long ago and preserved in its shape.

That is the first layer.

Then comes the sample.

A piece of rock is removed, labeled, and carried away from the field. In the laboratory, it is cut and polished until light can pass through it. Under magnification, the surface resolves into structure.

Mineral grains appear. Boundaries sharpen. Patterns that were invisible in the field become clear.

Some minerals grow aligned, reflecting the direction of stress. Others have been altered, partially melted and recrystallized under changing conditions. What looked uniform reveals a history.

Then the scale shifts again.

Individual minerals are separated and tested. Isotopes are measured. Ratios are calculated with precision that does not exist in the field.

From that, a number emerges.

1.9 billion years.

2.1 billion years.

A moment in deep time, carried within the structure of the rock.

One sample gives one point.

Then another is tested.

And another.

Across a region, those points begin to cluster. The timelines begin to overlap. What seemed like isolated rocks start to define a shared interval, a phase of formation, a period of deformation.

The pattern expands.

In another part of the world, separated by an ocean, similar work is being done. A different outcrop. A different landscape. But the process is the same.

Measurements are taken.

Samples are collected.

Ages are determined.

At first, there is no reason to connect the two regions.

They exist in different continents, under different conditions, within entirely separate modern systems.

But the numbers begin to converge.

The timelines begin to overlap within a narrow window. The deformation follows the same direction. Layers are folded in comparable ways. Minerals record similar pressures and temperatures.

The resemblance is rarely exact.

It never is.

But it is close enough to suggest something more than coincidence.

So the comparison deepens.

Structures are mapped more carefully. Orientations are analyzed. The direction of compression is reconstructed from the alignment of minerals and the geometry of folds. What once appeared local begins to extend beyond its immediate setting.

The possibility of connection takes shape.

Then another layer emerges.

In the same regions, mineral deposits begin to appear. At first, they are identified through surface exposure or small-scale exploration. A vein of quartz carrying traces of gold. A zone enriched in copper. Iron formations layered within the rock.

Individually, each deposit appears local.

But as mapping continues, the boundaries expand.

Additional zones are identified. The deposits begin to fall into the same zones, following the same structural pathways. A belt forms, a region where specific elements are concentrated in a way that reflects the same underlying processes.

That belt does not stop at the edge of a modern continent.

It simply disappears beneath younger rock or beneath the ocean.

Until it is found again elsewhere.

In another region, another team identifies a similar system. The same types of deposits. The same structural setting. The same relative position within the crust.

When these regions are compared, the alignment becomes difficult to ignore.

What appears scattered begins to resolve into continuity.

A system that once extended across a larger landmass is now broken into fragments, each preserved within a different continent.

Not perfectly.

But clearly enough.

Still, the connection is not assumed.

It is tested.

Data from multiple regions is brought together. Ages, structures, mineral systems, each adds a constraint. Models are built, not as fixed maps, but as possibilities. Fragments are shifted, rotated, and aligned in different configurations.

Some arrangements fail quickly.

Structures do not match. Timelines conflict. Patterns break apart under closer examination.

Others hold.

They satisfy multiple conditions at once. The structures align. The timelines overlap. The deposits fall into consistent zones.

The fit is not exact.

But it is consistent.

And consistency, across independent lines of evidence, carries weight.

Even then, the process is not final.

New data can change the picture. A sample analyzed with greater precision may shift an age. A newly mapped structure may extend a boundary. A deposit once thought isolated may connect to a larger system.

The reconstruction adjusts.

It becomes more refined, not by becoming complete, but by becoming better constrained.

There are always gaps.

Some regions have been altered beyond recognition. Later tectonic events have overprinted earlier structures. Erosion has removed entire sections of the record. In many places, the evidence lies buried beneath younger rock, inaccessible to direct observation.

The past is not preserved evenly.

It is fragmented.

But it is not lost.

Enough remains to trace the outlines.

Across continents, the same patterns appear again and again. Similar timelines. Comparable structures. Recurring mineral zones. Each on its own is limited. Together, they begin to overlap.

And in that overlap, the connection emerges.

The continents are not isolated.

They are remnants.

Pieces of a larger system that once existed as a single structure, assembled, reshaped, and later broken apart.

What is seen today is the result of that separation.

What is found in the rock is the record of what came before.

The work of reconstruction does not reveal a perfect image.

It reveals a structure that becomes clearer as more pieces are added.

And those pieces are still being found.

Not in one place.

But everywhere the ancient crust remains exposed, waiting to be measured, sampled, and connected.

The surface may hide the past.

The rock does not.

Chapter 12
Following the Ancient Scars

Finding the clues is only the beginning.

By this point, fragments of evidence exist across the continents, matching ages, aligned structures, shared mineral systems. Individually, each is meaningful. Together, they suggest connection.

But suggestion is not enough.

The challenge is not locating the pieces.

It is determining how they fit, and just as importantly, how they do not.

The work moves beyond observation into constraint.

Field mapping provides the first layer, structure, orientation, and boundaries traced across exposed rock. Laboratory analysis adds time and composition, establishing when events occurred and how materials changed. Geophysical data extends the view downward, revealing variations in density and structure that cannot be seen at the surface.

Each layer contributes information.

None is complete on its own.

The process becomes one of integration.

Data from a single region is assembled into a coherent picture. Structure shows how rock was deformed. Chemistry reveals how it evolved. Age defines when those changes took place. Together, they describe a system, not in isolation, but as part of a broader history.

Then the comparison begins.

Information from one region is tested against another, often across vast distances. A structure mapped in one location is examined alongside similar features elsewhere. Ages are compared. Mineral systems are evaluated within their structural context. Geophysical signatures are analyzed for continuity at depth.

At first, the connections are tentative.

A similar age. A comparable structure.

But similarity alone does not establish connection.

The test is whether multiple lines of evidence converge.

To explore this, continents are repositioned.

Fragments are rotated, shifted, and aligned in different configurations. Each arrangement is evaluated against the available data. The goal is not to find a single exact fit, but to determine which configurations remain viable under constraint.

Most do not.

Structures fail to align. Timelines diverge beyond acceptable ranges. Mineral systems fall out of position. Patterns that appear convincing at one scale break apart when examined more closely.

Failure is common, and necessary.

Each rejected configuration narrows the possibilities. Each inconsistency removes a potential arrangement, refining the range of what remains plausible.

What survives is not perfectly constrained.

But it is consistent.

In successful models, multiple constraints are satisfied at once. Structures align across regions. Timelines overlap within defined limits. Mineral systems fall into coherent zones. Geophysical data supports the arrangement at depth.

No single line of evidence carries the reconstruction.

It is the convergence that matters.

Even then, the model is not fixed.

New data can alter the picture. A more precise age may shift the sequence of events. A newly identified structure may extend or redefine a boundary. Improved imaging may reveal features that were previously hidden.

Each addition requires reevaluation.

The reconstruction adjusts, not by becoming final, but by becoming better constrained.

To see how this works, consider a single boundary.

In one region, a suture zone marks where two fragments once collided. Its structure is mapped. Its age is established. Its associated mineral systems are identified.

Across an ocean, a similar zone is recognized.

At first, the connection is uncertain.

But when tested within a model, the features begin to align. The deformation follows the same direction. The timelines fall within a comparable range. Mineral systems occupy similar positions relative to the boundary. Geophysical data suggests continuity below the surface.

The match is not exact in detail.

It does not need to be.

It needs to be consistent across independent lines of evidence.

When that consistency holds, the connection strengthens.

When it does not, the model is revised.

This process repeats across multiple regions.

Each successful alignment adds clarity. Each rejected configuration sharpens the constraints. Gradually, the scattered fragments begin to resolve into a coherent structure.

Not a perfect map.

But a working system.

There are limits.

Some regions have been altered beyond recognition. Later tectonic events have overwritten earlier structures. Erosion has removed vast portions of the record. In many areas, the evidence remains buried beneath younger rock, inaccessible to direct observation.

Uncertainty is not a flaw.

It is part of the reconstruction.

What matters is not completeness, but reliability within the available constraints.

The continents preserve the scars of ancient collisions, suture zones, metamorphic belts, and mineral systems that mark where fragments once joined. By following these scars, geologists trace the outlines of a world that no longer exists.

The picture that emerges is not fixed.

It is refined over time.

Each new observation has the potential to adjust the model, to tighten the constraints, or to reveal connections that were previously unseen.

The continents have changed.

The record has not.

It remains embedded within the crust, waiting not just to be found, but to be tested, compared, and connected.

Chapter 13
Where the Next Discoveries May Lie

The outline of Columbia is clear enough to be recognized.

Its fragments can be traced. Its assembly and breakup can be described. The major structures that defined it, cratons, suture zones, mineral belts, are known.

But the picture is not complete, and never will be.

What exists today is a working reconstruction, one built from evidence, refined over time, and open to revision as new data emerges. Some regions are well understood. Others remain uncertain. Entire sections of ancient crust still hold information that has not yet been recognized or fully interpreted.

The next discoveries will not come from a single place.

They will come from many.

Some begin in the field.

In remote regions, where ancient crust is exposed at the surface, mapping continues. In northern Canada, a team moves slowly across a stretch of shield rock, marking boundaries between units that appear nearly identical at first glance. A subtle change in texture, a shift in mineral alignment, a variation in structure, small differences that suggest a boundary.

That boundary matters.

It may represent the edge of a fragment. A transition between two crustal blocks. A piece of a larger system that has not yet been fully traced.

The work is careful.

Day after day, observations are added to a growing map. What begins as isolated measurements becomes a pattern. Lines extend. Regions connect. A structure that once seemed local begins to take on broader significance.

The discovery is not sudden.

It builds over time.

Other discoveries begin far from the field.

In a laboratory, a mineral sample is analyzed with increasing precision. Techniques that once produced approximate ages now resolve differences of tens of millions of years. A rock previously assigned to a single event is revealed to record multiple stages, formation, deformation, alteration.

The timeline shifts.

What was once considered one collision may become two. What appeared to be a single phase of growth may resolve into a sequence of events, each contributing to the assembly of the supercontinent.

The structure remains.

But the history becomes more detailed.

Technology continues to extend what can be seen.

Geophysical surveys now cover larger areas with higher resolution. Data collected from satellites, aircraft, and ground-based instruments reveal subtle variations in gravity and magnetism across entire continents. Patterns emerge that do not correspond to surface features but reflect differences in composition and structure at depth.

In one region, a faint linear anomaly appears in the data.

At first, it is unclear what it represents.

But as more information is gathered, the anomaly aligns with structures mapped at the surface. It extends beyond them, continuing beneath younger rock where direct observation is not possible.

A hidden boundary is revealed.

It may represent an extension of a suture zone, evidence of a collision that was only partially understood before.

What was invisible becomes part of the map.

Mineral exploration offers another pathway.

A drilling program begins in a region with known potential. Core samples are brought to the surface, revealing structures that were never exposed. Layers appear that do not match expectations. Mineralization occurs in patterns that suggest a broader system.

The deposit is studied.

Its structure is mapped. Its age is determined. Its relationship to surrounding rock is analyzed. As the data accumulates, connections begin to emerge.

The deposit is not isolated.

It fits into a larger pattern.

When compared with data from other regions, it aligns with known systems, extending a mineral belt farther than previously recognized, linking fragments across continents.

Economic activity leads to geological insight.

Even well-studied regions continue to yield new information.

Old maps are revisited. Samples collected decades earlier are reanalyzed with improved techniques. Data that once seemed incomplete or inconsistent is reinterpreted within a new framework.

A structure mapped years ago may take on new meaning.

A boundary once dismissed as local may become part of a larger connection.

The past is not fixed.

It becomes clearer as the tools improve.

The process is cumulative.

Each new observation, whether from fieldwork, laboratory analysis, or geophysical imaging, adds to a growing network of evidence. Over time, that network becomes more detailed, more constrained, and more consistent.

But it does not become complete.

There are limits to what can be recovered.

The Earth's surface has been reshaped repeatedly over billions of years. Erosion has removed vast amounts of material. Later tectonic events have altered or destroyed earlier structures. Some evidence lies buried beyond reach. Some has been erased entirely.

Reconstruction is always partial.

Even so, the broader structure remains.

Columbia existed. It formed through the assembly of ancient cratons, persisted for hundreds of millions of years, and eventually broke apart. Its fragments now lie within modern continents, preserving the record of that history.

Future discoveries are unlikely to overturn that conclusion.

They will refine it.

They will adjust the positions of fragments, clarify the sequence of events, and reveal details that are not yet visible. They will make the picture sharper, even if it can never be complete.

The search continues because the record remains.

Every outcrop, every mineral deposit, every geophysical signal carries information about the past. Each has the potential to connect with others, forming patterns that extend beyond any single location.

The continents are not silent.

They are still revealing their history.

And as new discoveries are made, the reconstruction of Columbia will continue to evolve, becoming not a finished map, but an increasingly detailed understanding of how one of Earth's earliest supercontinents came to be.

Chapter 14
The Supercontinent Cycle

Columbia was not a one-time event.

It was part of a pattern, one that has shaped the Earth for billions of years and continues to operate today. Continents do not simply form and remain in place. They move, interact, assemble, and break apart in a repeating sequence.

This pattern is known as the supercontinent cycle (see Figure 14.1).

At any moment, the surface of the Earth is in continuous motion.

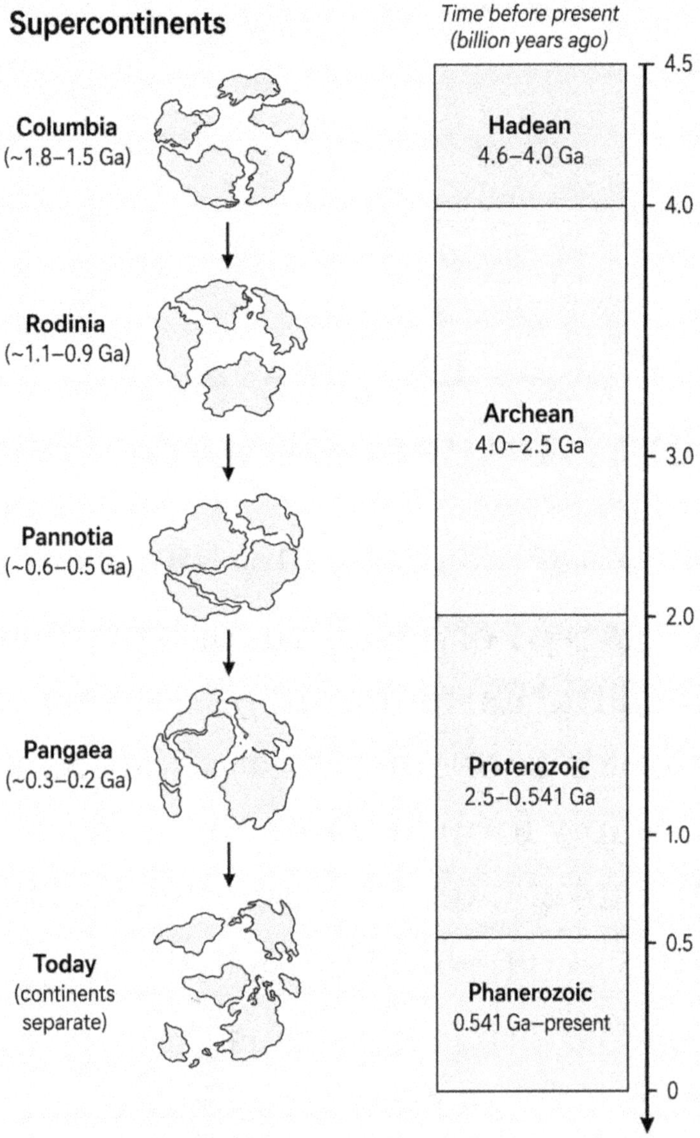

Figure 14.1 — Geological timeline

The movement is slow. A continent may shift only a few centimeters in a year. Standing in one place, nothing appears to change. The ground feels fixed. The horizon remains the same.

But over time, the motion accumulates.

Over one million years, a continent moves the length of a city.

Over ten million years, it crosses a region.

Over one hundred million years, it can traverse an ocean basin.

What feels still is moving.

The cycle begins with separation.

Fragments of continental crust drift across the planet, carried by tectonic plates moving over the mantle below. Oceans lie between them, expanding where plates pull apart and new crust forms along mid-ocean ridges.

For long periods, the fragments remain apart.

But the system does not remain balanced.

Elsewhere on the planet, oceanic crust begins to sink beneath continental margins. Subduction zones form. Ocean basins slowly begin to close. The distance between continents decreases, not everywhere at once, but in specific regions where the geometry of plate motion brings them together.

The process is uneven.

One ocean may be expanding while another is shrinking.

Over tens of millions of years, the shrinking ocean disappears.

The continents approach.

At first, the boundary between them is marked by volcanic arcs, long chains of volcanoes forming above subduction zones. The ocean floor between the continents is consumed. Sediment accumulates along the margins. The system tightens as the distance continues to close.

Then the continents meet.

The collision is not sudden.

It unfolds over millions of years as the buoyant continental crust resists being pulled downward. Instead of sinking, it thickens. Rock layers fold and compress. Large sections of crust are pushed over one another.

Mountains rise.

They do not appear overnight. They grow slowly, building height as the crust thickens beneath them. Entire regions are uplifted. Landscapes are reshaped.

At this stage, the system is dominated by convergence.

Fragment after fragment joins the growing landmass. Each collision adds complexity. Each creates new mountain belts, new suture zones, new structures within the crust.

Eventually, a supercontinent forms.

For a time, it holds.

The interior becomes relatively stable compared to the active margins. The landmass stretches across vast distances. Climate patterns shift in response to its size and configuration. Ocean circulation changes. The planet adjusts to the presence of a single, dominant continent.

But the stability is temporary.

Beneath the surface, heat continues to rise.

The thick crust of the supercontinent acts as a barrier, slowing the loss of heat from the mantle below. Over tens of millions of years, that heat builds. The base of the crust warms. Rock begins to weaken.

The system changes from below.

Small zones of extension appear within the interior. These are not yet breaks, but weaknesses, regions where the crust is beginning to stretch.

Rifts begin to form.

At the surface, they may appear as long, narrow valleys. At depth, the crust thins. Mantle material rises. Magma forms and moves into the fractures.

The process feeds itself.

As the crust stretches, it weakens. As it weakens, it stretches more easily.

If the process continues, the continent begins to split.

A new boundary forms.

Magma rises continuously along this boundary, creating new oceanic crust as the two sides move apart. What was once a single landmass separates into fragments.

A new ocean is born.

The supercontinent breaks apart.

The fragments drift away, carried by tectonic motion. Coastlines form. Oceans expand. The configuration of the planet changes again.

The system returns to a state of separation.

But not indefinitely.

As the fragments move, new subduction zones form. Ocean basins begin to close in other regions. The cycle shifts once more toward convergence.

Over hundreds of millions of years, the process repeats.

Columbia followed this path.

It formed through the convergence of ancient cratons. It stabilized as a connected landmass. Heat accumulated beneath it. Rifting began. It broke apart, scattering its fragments across the globe.

Those fragments did not remain separate.

They moved.

Over time, many became part of another supercontinent: Rodinia. That configuration persisted, then broke apart, repeating the same sequence. Much later, the continents assembled again into

Pangaea, whose breakup shaped the arrangement we recognize today.

Each cycle is similar.

But not identical.

The positions of continents change. The timing varies. Some fragments remain intact through multiple cycles, acting as stable cores. Others are reshaped, divided, or recombined in new ways.

The Earth does not return to a previous state.

It builds on what came before.

Structures formed during one cycle influence the next. Ancient suture zones may become zones of weakness during later rifting. Regions of thickened crust may guide future deformation. Mineral systems formed during earlier collisions remain embedded within the crust, carried forward through time.

The past shapes the future.

The cycle continues today.

The Atlantic Ocean is widening as the Americas move away from Europe and Africa. At the same time, the Pacific basin is shrinking, with subduction zones consuming oceanic crust along its margins.

These opposing motions are part of the same system.

They suggest a future in which the continents will once again come together.

The timescale is immense.

Over the next 200 to 300 million years, the current configuration will change. Oceans will continue to open and close. Fragments will move across the planet. Eventually, a new supercontinent will form.

It will not look like Columbia, Rodinia, or Pangaea.

But it will emerge from the same processes, shaped by the same forces, and influenced by structures that have persisted through deep time.

The cycle continues, and it will repeat.

Understanding this cycle changes how we see the Earth.

The continents are not fixed. The oceans are not permanent. The surface of the planet is part of an ongoing system, one that has been unfolding for billions of years and will continue long into the future.

Columbia was one phase of that system.

Its formation and breakup were not isolated events, but part of a continuous process that links the past, present, and future of the Earth.

In the final chapter, we return to what remains, how the record of that ancient world is preserved, and what it reveals when the fragments are brought together.

Chapter 15
Treasures Written in Stone

The supercontinent Columbia no longer exists.

Its mountains have been worn down. Its coastlines have vanished. The landmass itself has been broken apart and carried across the globe, its fragments embedded within the continents we know today.

Nothing of it remains in the form it once held.

What remains is the record.

That record is written in stone.

Across the continents, ancient rocks preserve the story of Columbia's formation and breakup. These rocks do not present the past as a single, complete picture. They exist as fragments, scattered, altered, and separated by immense distances.

At first glance, they appear unrelated.

A band of deformed rock in one region. A mineral deposit in another. A cluster of ancient ages preserved across distant continents. Each seems to describe only its immediate surroundings.

But when these fragments are compared, patterns begin to emerge.

Cratons mark the oldest surviving pieces of continental crust. Their deep roots preserve conditions that have remained stable for billions of years. Metamorphic belts trace the remains of mountain systems that no longer exist. Suture zones mark the boundaries where continents once collided. Mineral deposits reveal the movement of heat, magma, and fluids through the crust.

Each piece holds part of the story.

None tells it alone.

To reconstruct Columbia, geologists bring these fragments together. They compare ages. They align structures. They trace mineral systems across continents. They use geophysical data to extend what is known at the surface into the depths below.

The process is not direct.

It does not involve observing the past, but interpreting the record that remains. Each rock carries information about the conditions under which it formed, pressure, temperature, chemical environment, and time.

These are not visible qualities.

They are encoded in structure and composition.

Reading that record requires careful analysis, comparison, and interpretation. It requires connecting pieces that are no longer physically joined. It requires building a coherent picture from incomplete information.

The Earth does not present its history clearly.

It preserves it.

Even as the surface changes, the deeper record endures.

The mountains of Columbia are gone, but their roots remain. The oceans that once separated continents have vanished, but their boundaries are preserved in suture zones. The processes that built the supercontinent continue to shape the distribution of materials within the crust.

The past is not visible.

But it is present.

It is embedded within the structure of the continents, carried forward through time as the surface of the planet evolves.

Columbia was one of the earliest large-scale expressions of continental assembly. Its formation marked a shift in the development of the Earth's crust, a point at which fragments of ancient material came together to form a unified system on a planetary scale.

Its breakup did not erase that system.

It redistributed it.

The fragments that once formed Columbia became part of the foundation for later continents and future supercontinents. Structures created during its assembly influenced the way the crust behaved in later cycles. Mineral systems formed during its collisions remained preserved, becoming part of the resources that shape human activity today.

Its influence continues.

The continents we see now are not independent features. They are the result of repeated cycles of assembly and breakup, shaped by processes that have operated for billions of years.

Each carries the imprint of earlier worlds.

To understand the present, those earlier worlds must be reconstructed. Not as complete landscapes, but as patterns preserved within the Earth.

To understand the future, the pattern must be recognized.

Continents will continue to move. Oceans will open and close. New mountain ranges will rise, and old ones will erode away. The configuration of the planet will change again and again.

The cycle will continue.

But each cycle leaves a record.

That record accumulates.

It is not erased when continents break apart. It is carried forward, embedded within the crust, influencing what comes next. The Earth evolves by building on its past, not by replacing it.

Columbia is part of that accumulation.

It is not visible in the way modern continents are visible. It cannot be pointed to on a map. But it exists within the structure of the Earth, preserved across billions of years.

Its presence can be traced.

Its history can be read.

And its influence can still be seen in the arrangement of continents and the distribution of materials across the planet.

The surface of the Earth is temporary.

The deeper record continues to endure.

Stand on a stretch of exposed rock in northern Canada, or in the quiet expanses of West Africa, or on the ancient ground of Australia.

At first, it feels unremarkable.

Flat stone. Scattered lakes. Low ridges shaped by erosion.

Nothing that suggests a world in motion.

But beneath that surface lies a history measured in billions of years.

The rock beneath your feet has been buried deep within the Earth, altered under immense pressure, lifted, eroded, and carried across the planet. It has been part of mountains that no longer exist, of continents that have assembled and broken apart, of systems that reshaped the surface of the Earth again and again.

It has been part of Columbia.

What appears still is not still.

What appears simple is not simple.

It is the surface of a planet that has never stopped changing.

And within that surface, the record remains, written in stone, carried through time, waiting to be read.

www.ingramcontent.com/pod-product-compliance
Lightning Source LLC
LaVergne TN
LVHW050820061126
851857LV00012B/893
9798349509056